CURIOUS MINDS

A short compilation of life and work of ten Curious Minds

Volume-I

HIMANGSU SARMA

ISBN:9798858249214

DEDICATION

I dedicate this book to my father Lt. Ananta Sarma, my mother Mrs. Bina Sarma and all my teachers.

Contents

Preface

Welcome to the captivating world of "Curious Minds" This book is a tribute to the boundless human spirit of inquiry and discovery that has led to some of the most profound advancements in science and our understanding of the universe.

The journey of scientific exploration is a tapestry woven with the threads of curiosity, perseverance, and imagination. Within these pages, we embark on an expedition through time, venturing into the lives of brilliant minds who dared to challenge conventional wisdom and reshape the contours of knowledge. From ancient philosophers who contemplated the nature of the cosmos to modern trailblazers unraveling the mysteries of quantum mechanics, this book celebrates the breadth and depth of scientific inquiry.

While the contributions of the scientists showcased here are vast and varied, they share a common thread: a relentless pursuit of truth. With determination that transcended challenges and setbacks, these individuals have illuminated the darkness of the unknown, one discovery at a time. Their stories serve as beacons of inspiration for all those who seek to understand the world and contribute to its betterment.

"Curious Minds" is not just a chronicle of achievements; it's an exploration of the human spirit's undying desire to push the boundaries of knowledge. Whether you are an aspiring scientist, a curious mind, or someone intrigued by the wonders of the universe, this book offers a front-row seat to the incredible journeys of those who shaped our understanding of reality.

As you immerse yourself in these stories, I invite you to reflect on the indomitable courage it takes to challenge the unknown, the passion required to delve into uncharted territories, and the resilience needed to rise above failures. May these narratives kindle your curiosity, spark your creativity, and ignite a lifelong appreciation for the beauty and complexity of the scientific endeavor.

In the words of Albert Einstein, "The important thing is not to stop questioning. Curiosity has its own reason for existence." Let us journey together through the lives of the owners of these brilliant minds and may their stories inspire you to question, to explore, and to forge your own path of discovery.

Himangsu Sarma

20/08/2023

Acknowledgments

In the creation of 'Curious Minds' there are hearts and hands that have played an indispensable role in shaping its pages and bringing its vision to life. To these individuals, I extend my deepest gratitude and heartfelt appreciation.

To the most treasured piece of my heart, Ahaan, for drawing my attention to create books for children. Without you in my life, I probably wouldn't ever be cognizant of my desire to walk in this path of writing books for young and brilliant humans like you.

To my beloved wife, Dr. Rijusmita Sarma, your unwavering support, patience and belief in my endeavours have been my guiding light. Your encouragement from the inception of the idea to the late-night writing sessions and your understanding of the time devoted to this project have truly been my anchor. You have been with me throughout the process of writing, editing and designing with your suggestions and feedback. This book is a testament to the partnership that fuels my aspirations and the love that enriches every facet of my life.

To my sisters, you have been a constant source of encouragement in the journey of life. Your unending enthusiasm, feedback and invaluable insights have been instrumental. Your belief in my capabilities has been a source of strength.

To my esteemed colleagues, your collaboration and dedication have been a driving force behind the completion of this project. The insightful discussions with you have enhanced the depth of the content. The exchange of ideas and the shared passion for knowledge have truly enriched this endeavour and I am grateful for your presence on this voyage.

To all those people whose names may not have graced these pages but whose support has been felt in every keystroke and paragraph, I offer my heartfelt appreciation. From the friends who offered words of encouragement to the mentors who guided my path, this book carries a piece of the collective spirit that has propelled it forward.

In the end, a project of this nature is a symphony of contributions from many, and I am humbled by the orchestra that has played its part. With gratitude and a sense of accomplishment, I present 'Curious Minds' as a product of shared inspiration and collaborative effort.

Thank you all for your roles in making this vision a reality.

Warm regards,

Himangsu Sarma
20/08/2023

Chapter 1

Galileo's Starry Quest

In the bustling streets of Pisa, there lived a young boy named Galileo. He had a twinkle in his eye and an unquenchable curiosity for the world around him. From a very young age, Galileo loved observing the stars that adorned the night sky. He would lie on the grass in his backyard, gazing up at the twinkling lights and dreaming of the mysteries they held.

Galileo's father hoped he would become a doctor like himself but Galileo's heart was set on understanding the heavens. He would sneak away from his studies and spend hours at the local observatory, peering through telescopes and sketching what he saw.

One evening, as Galileo focused his telescope on the Moon, he made a discovery that would change his life forever. Instead of the smooth, flawless surface everyone believed the Moon had, he saw mountains and craters! Galileo realized that the Moon was a world just like ours with its own rugged landscape.

Excited by this revelation, Galileo pointed his telescope towards the night's brightest star, Venus. To his amazement, he saw that Venus went through phases, just like the Moon. It was as if Venus had its own set of changing faces. This observation provided strong evidence that planets orbited the Sun, not the Earth as everyone thought.

Galileo's discoveries didn't stop there. He observed Jupiter and its four largest moons, which he named the Galilean moons. This challenged the belief that everything in the universe revolved around Earth. He saw the rings of Saturn, though he couldn't quite make out their true shape. And he even turned his telescope towards the Sun, discovering sunspots and confirming that the Sun wasn't a perfect, unchanging sphere.

As word of Galileo's observations spread, some people were fascinated by his work, while others were threatened by the new ideas he brought. The established beliefs of the time clashed with Galileo's discoveries, and he faced challenges from those who feared change.

Galileo's determination, however, couldn't be swayed. He continued his observations and experiments, even in the face of adversity. He knew that understanding the universe was worth the struggles he faced.

Eventually, Galileo's findings led him to support the idea put forth by Copernicus: that the Earth orbited the Sun. This was a radical idea that challenged the widely accepted geocentric model. Galileo's support of this theory got him into trouble with the authorities, who accused him of heresy. He was put on trial and forced to recant his beliefs.

But even in his later years, when he was put under house arrest, Galileo continued to explore and write about the wonders of the universe. His books and ideas inspired future generations of scientists to question, observe, and explore.

Galileo's story reminds us that curiosity and the pursuit of knowledge can lead to amazing discoveries, even in the face of challenges. He showed us that sometimes, the greatest adventures are found by looking up at the stars and daring to ask, "Why?

Isaac and the Magical Apple

Once upon a time, in a quaint village nestled between rolling hills, lived a young boy named Isaac. He was known for his insatiable curiosity and boundless imagination. Isaac had an insistent urge to uncover the mysteries of the world around him. Every day, he would explore the forests, gaze at the stars, and question everything he encountered.

Isaac's favourite place was under the large apple tree behind his house. It was there, one sunny afternoon, that a falling apple caught his attention. "Why did the apple fall straight to the ground?" Isaac wondered. This simple question marked the beginning of his incredible journey of discovery.

Isaac devoted himself to understanding the forces of nature. He read books, conducted experiments, and observed the world around him. His tiny room soon became a haven for his experiments - from pendulums swinging back and forth to water being poured into different-shaped containers. Every discovery made Isaac's heart dance with joy.

One day, as Isaac pondered the concept of gravity, a brilliant idea struck him like a lightning bolt. He had an epiphany - what if the same force that caused the apple to fall was responsible for keeping the Moon in its orbit around Earth? This revelation ignited a fire within Isaac to explore this idea further.

His dedication led him to develop mathematical equations that described the motion of objects under the influence of gravity. He called these equations "laws of motion," and they would change the way the world understood the universe.

Isaac's discoveries didn't stop there. He delved into the study of light and colour, conducting experiments with prisms and rainbows. He uncovered that white light was actually a mixture of different colours, and this revelation was the birth of the field of optics.

Word of Isaac's groundbreaking work spread far and wide. Scholars and scientists from across the world travelled to meet him, amazed by his brilliance and dedication. Isaac's name became synonymous with knowledge and innovation.

But Isaac remained humble and focused on his quest for understanding. He continued to make discoveries, each one building upon the last, as he explored the mysteries of the universe.

As Isaac grew older, he shared his knowledge with others, teaching at universities and publishing his findings. His legacy lived on through his teachings and writings, inspiring generations of scientists to come.

And so, the curious boy who questioned the falling apple became one of history's greatest scientists, unlocking secrets of the universe through his insatiable curiosity and boundless determination. Isaac's story teaches us that asking questions and seeking answers can lead to incredible discoveries, forever changing the way we understand the world around us.

Chapter 3

Thomas and the Bright Ideas

In the bustling town of Milan, Ohio, there lived a young boy named Thomas. Thomas was an inquisitive soul with a heart full of curiosity. He was always up to something - whether it was taking things apart to see how they worked or dreaming up new inventions.

Thomas's love for exploring the world

around him started at an early age. He loved to spend hours in the woods, observing the plants and creatures that called it home. He would often sit by the river, watching the water flow and imagining the mysteries it held.

One day, as Thomas was walking by the tracks, he saw a bird's nest perched atop a tall pole. Curious as ever, he climbed the pole to get a closer look. But as he climbed, he accidentally slipped and fell onto the tracks below. A kind station agent pulled him to safety just in time, but young Thomas was shaken by the close call.

This incident sparked something within Thomas. He became determined to make the world a safer place. He began to think about ways to improve the things around him, to make life better for everyone.

Thomas's first big invention was a machine that used telegraph signals to report news. This invention caught the attention of people in the nearby town of Detroit, where he moved to work on improving telegraph systems. But Thomas's mind was already spinning with new ideas.

He dreamed of creating something that could record and replay sound. He wanted to capture music, people's voices, and the sounds of the world. After countless experiments, Thomas invented the phonograph, a device that could record and play back sounds. It was a magical creation that brought music and voices to life, delighting people all around the world.

But Thomas's journey didn't stop there. He wanted to bring light to every home, even when the sun had set. He spent years experimenting with different materials until he finally discovered the perfect filament for his electric light bulb. It was a moment of triumph when Thomas's bulb lit up, illuminating the darkness and paving the way for modern lighting.

Thomas Edison's inventions continued to

change the world, leaving his mark on history as one of the greatest inventors of all time.

Thomas's story teaches us that curiosity, determination and a willingness to learn from mistakes which can lead to amazing discoveries. He showed that even when things seem tough, with hard work and a bit of imagination, we can bring light to the darkest corners and make the world a better place. So, the next time you flip a light switch, remember the boy from Milan who brightened the world with his ideas.

Chapter 4

The Magical Voice of Alexander Graham Bell

In the quaint town of Edinburgh, Scotland, there lived a curious boy named Alexander. From a young age, he was captivated by the world around him. He loved watching his father, who was a speech therapist, work with people who had trouble speaking. Little did he know that this fascination would lead him on a remarkable journey of discovery.

Alexander's mother was deaf and he often communicated with her using simple signs and gestures. But he dreamed of finding a way for her to hear the sounds of the world. He believed that if he could do that, he would help not only his mother but many others as well.

As he grew older, Alexander's curiosity and determination only grew stronger. He loved experimenting and inventing things. He even built a machine to help his dog, Trouve, climb up the stairs!

One day, while experimenting with sound vibrations, Alexander had a breakthrough. He realized that if he could send sound vibrations through a wire, it might be possible to transmit voices over long distances. This idea was like magic to him and he was determined to make it a reality.

With the help of his loyal assistant, Thomas Watson, Alexander worked tirelessly on his invention. Countless hours were spent trying different materials and configurations. One fateful day, as Alexander spilled some acid, he accidentally spilled some on a piece of metal. A sound came through the wire, and Thomas heard Alexander's voice! "Mr. Watson, come here, I want to see you," Alexander exclaimed. And in that moment, the telephone was born.

News of Alexander's incredible invention spread like wildfire. People were amazed by the idea of talking to someone far away, as if they were right beside you. Soon, telephones were popping up everywhere, connecting people across cities and even countries.

But Alexander's brilliance didn't stop there. He was always eager to learn and discover new things. He worked on improving the telegraph, developed the metal detector and even investigated ways to help people with speech difficulties communicate more easily.

Alexander's heart was as big as his inventions. He used his fame and fortune to support causes he believed in, including helping people with hearing impairments. He believed that everyone should have a chance to communicate and be heard.

The magical voice of Alexander Graham Bell continued to echo through history. His inventions transformed the way we communicate and brought people closer together. He showed the world that a curious mind, a dream and hard work could create wonders beyond imagination.

So, the next time you pick up a phone to chat with a friend or hear someone's voice on the other end, remember the boy from Edinburgh who turned his dreams into reality. Alexander Graham Bell's story reminds us that with passion and determination, even the wildest ideas can come to life.

Chapter 5

Aryabhatta's Astral Adventures

In the ancient kingdom of Magadha, nestled along the banks of the Ganges River, lived a young and inquisitive boy named Aryabhatta. From a tender age, Aryabhatta's eyes glistened with wonder as he observed the celestial dance of the stars and planets in the night sky. Little did he know that his curiosity would lead him on a journey of discovery that would forever change the world of mathematics and astronomy.

Aryabhatta's fascination with the universe began when he noticed the patterns in the stars and their movement across the sky. He spent countless nights studying the heavens, trying to make sense of the mysteries they held. He dreamed of understanding the universe's intricate design and the mathematical secrets hidden within.

As he grew older, Aryabhatta's curiosity only deepened. He dedicated himself to the study of mathematics and astronomy, eager to unravel the cosmic puzzle. He marvelled at the cycles of the moon, the positions of the planets and the rhythm of the seasons.

One day, as Aryabhatta sat under a tree, he had an epiphany. He realized that the Earth wasn't flat as many believed, but rather a sphere that revolved around the Sun. This groundbreaking insight laid the foundation for his revolutionary work in astronomy.

Aryabhatta's greatest achievement was his work "Aryabhatiya," a brilliant treatise on mathematics and astronomy. In this work, he introduced the concept of "zero," a numeral that holds immense significance in modern mathematics. He developed algorithms for calculations and provided precise explanations for the movements of celestial bodies.

His contributions to trigonometry, arithmetic, and algebra were not only ahead of his time but also paved the way for future mathematicians and astronomers. Aryabhatta's theories and formulas transformed the way people understood the cosmos and mathematics, inspiring generations to explore the mysteries of the universe.

Aryabhatta's legacy lived on as his work spread across the Indian subcontinent and beyond. His discoveries influenced scholars and scientists for centuries, shaping the course of mathematics and astronomy. He showed the world that a curious mind, a dedication to learning and a passion for discovery could lead to incredible achievements.

As you gaze at the stars in the night sky or work on solving mathematical puzzles, remember the young boy from Magadha who dared to question, explore and dream. Aryabhatta's astral adventures remind us that the universe is full of mysteries waiting to be uncovered and our curiosity can guide us to new and exciting discoveries.

Chapter 6

Marie's Radiant Discoveries

In a cozy town nestled between rolling hills, lived a curious and determined girl named Marie. She had a sparkle in her eyes that matched the stars in the night sky. Marie was always eager to learn and explore, especially the mysteries of the world hidden beneath the surface.

From a young age, Marie had a deep love for

science. She marvelled at the wonders of nature and dreamed of unlocking its secrets. She loved to ask questions and experiment, even if it meant getting a little messy. Her passion for knowledge set her on a path to change the world in ways she could never have imagined.

As Marie grew older, she attended university in Paris, where she continued to excel in her studies. It was there that she met Pierre Curie, a fellow scientist who shared her passion for discovery. They became partners in both science and life and their journey together would lead to some of the most remarkable breakthroughs in history.

Marie was particularly intrigued by a mysterious element called "uranium." She noticed that it emitted rays that could penetrate materials and even affect photographic plates. With Pierre's help, Marie discovered that uranium was not the only element with this property. They coined the term "radioactivity" to describe this phenomenon.

Marie and Pierre's dedication led them to the discovery of two new elements, polonium and radium, both of which were highly radioactive. Their groundbreaking research opened the door to a new understanding of the atom and its structure.

But Marie's journey wasn't without challenges. In a world where women were often overlooked in the scientific community, Marie faced discrimination and scepticism. However, her determination and brilliance shone through. She became the first woman to win a Nobel Prize, and later, she won a second Nobel Prize, this time in chemistry, for her work on radioactivity.

Marie's legacy extended beyond her

scientific achievements. She believed in using science for the betterment of humanity. During World War I, she established mobile radiography units, known as "Little Curies," to provide X-ray imaging for wounded soldiers. Her contributions saved countless lives and showcased the positive impact of science on society.

Marie Curie's story teaches us that passion, curiosity and determination can lead to incredible discoveries. She showed us that no dream is too big and that our efforts to understand the world can make a lasting impact. Marie's journey reminds us that even in the face of challenges, our pursuit of knowledge can change the world for the better.

So, the next time you look up at the stars or learn about the wonders of the universe, remember the determined girl who followed her curiosity and illuminated the mysteries of atoms. Marie Curie's magical discoveries continue to inspire us all to explore, ask questions, and make a difference.

Chapter 7

Chandra's Colourful Quest

In the vibrant city of Tiruchirapalli, India, lived a boy named Chandra. The inquiring light in his eyes was impossible to miss. From a young age, he was fascinated by the colours and patterns that danced in the world around him.

Chandra loved to explore nature and was especially drawn to the colours of the sky and the shimmering reflections on water. He wondered why the sky appeared blue during the day and why the setting sun painted the horizon with hues of red and orange. Little did he know that his fascination with colours would lead him to unlock one of science's most captivating secrets.

As Chandra grew older, his curiosity led him to the world of science. He studied physics with great enthusiasm, eager to find answers to his questions about light and colour. His insatiable thirst for knowledge caught the attention of his teachers, who recognized his brilliance and encouraged his pursuits.

One day, while he was working in a laboratory filled with sunlight, Chandra observed something extraordinary. He noticed that when light passed through a liquid, a tiny fraction of it scattered in all directions. This phenomenon, known as the "Raman Effect," fascinated him. Chandra realized that by studying this scattering, he could gain insight into the composition of matter.

Chandra's groundbreaking discovery not only explained why the sky appeared blue but also opened up a whole new world of possibilities. He found that by analysing the colours of scattered light, scientists could learn about the molecules and atoms that make up different materials

Chandra's work earned him the Nobel Prize in Physics, making him the first Indian scientist to receive this prestigious award. His discovery of the Raman Effect had a profound impact on the fields of physics and chemistry, enabling scientists to explore the fundamental building blocks of matter in new and exciting ways.

Chandra's story teaches us that a curious mind and a passion for discovery can lead to remarkable breakthroughs. He showed us that even the simplest of questions can lead to profound insights that transform our understanding of the world. Chandra's colourful quest reminds us that science is an adventure waiting to be explored, and our curiosity can guide us to uncover the most dazzling secrets of the universe.

So, the next time you see a rainbow or watch the colours of the sky change, remember the young boy from Tiruchirapalli who followed his curiosity and uncovered the hidden beauty of light. Chandra's journey encourages us all to look at the world with wonder and embark on our own colourful quests of discovery.

Chapter 8

Archimedes and the Eureka Moment

In the bustling city of Syracuse in ancient Greece, lived a brilliant young boy named Archimedes. The allure of curiosity was evident in his eyes and his mind was filled with ideas that would change the world of science and mathematics. Archimedes was known for his insatiable thirst for knowledge and his boundless imagination.

Archimedes had a natural gift for solving puzzles and understanding how things worked. He would spend hours observing the world around him, pondering questions and seeking answers. His favourite place was by the sea, where he loved to watch waves and tides play along the shore.

One day, as Archimedes was taking a bath, he had a remarkable idea. He noticed that the water level rose as he entered the tub, and a thought occurred to him: Could he use this principle to measure the volume of irregularly shaped objects?

Excited by his revelation, Archimedes jumped out of the bath and ran through the streets of Syracuse, shouting "Eureka! Eureka!" which means "I've found it!" in Greek. He couldn't wait to test his idea and share his discovery with the world.

Archimedes began experimenting with different objects and liquids. He realized that the volume of an irregularly shaped object could indeed be determined by the amount of water it displaced when submerged. This principle became known as "Archimedes' principle," and it revolutionized the way people understood buoyancy and volume.

Archimedes' ingenious insights extended beyond water displacement. He made significant contributions to geometry, calculating the value of pi and developing formulas to calculate the areas and volumes of various shapes. He even designed remarkable machines for warfare and engineering.

One of Archimedes' most famous inventions was the "Archimedes screw," a simple yet effective device for lifting water. This invention allowed water to be raised from lower levels to higher levels, aiding in irrigation and construction.

Archimedes' story teaches us that curiosity and creative thinking can lead to groundbreaking discoveries. He showed us that even the simplest observations can spark innovative ideas that transform our understanding of the world. His "Eureka moment" reminds us that innovation can strike at any time and it's important to be open to new ideas.

So, the next time you take a bath or observe something unusual, remember the young boy from ancient Greece who turned a simple observation into a revolutionary discovery. Archimedes' curiosity and imaginative thinking continue to inspire us to explore, question, and innovate.

Chapter 9

Nikola's Electrifying Dreams

In a quiet village nestled in Croatia, lived a young boy named Nikola. His gaze gleamed with wonder and his heart thumped to the rhythm of his vivid imagination. From an early age, Nikola was captivated by the mysteries of the natural world, especially the magic of electricity.

Nikola loved to watch lightning streak across the sky during thunderstorms. He wondered if he could harness that electrifying energy and use it to power machines and light up the world. His dreams were as big as the sky, and he believed that one day he could make them come true.

As he grew older, Nikola's fascination with electricity deepened. He read books and studied diligently, eager to understand the science behind electric currents and magnetic fields. He knew that unlocking the secrets of electricity could change the way people lived and worked.

Nikola's journey led him to the United States, where he worked alongside inventors and scientists who shared his passion for innovation. He had an idea that would revolutionize the way electricity was transmitted—alternating current, or AC. Unlike direct current, AC could travel over long distances without losing much energy.

With determination and hard work, Nikola developed the technology for AC electricity transmission. He built power plants that generated electricity using AC and created systems that could transmit this electricity across vast distances. His inventions laid the foundation for modern electrical systems, lighting up cities and powering homes.

But Nikola's dreams didn't stop at electricity. He had a vision of a world connected by wireless communication. He imagined a way to transmit information and messages through the air, without the need for wires. He built the Wardenclyffe Tower, an ambitious project that aimed to create a global communication network powered by wireless technology.

Although the tower was never completed due to financial challenges, Nikola's vision and contributions to electrical engineering left an indelible mark on the world. His inventions and ideas paved the way for the technologies we rely on today, from the electricity that powers our homes to the wireless communication that connects us.

Nikola Tesla's story teaches us that imagination and determination can lead to groundbreaking discoveries. He showed us that pursuing our dreams can result in transformative advancements that benefit humanity. His legacy as an inventor and visionary continues to inspire generations to think boldly and innovate.

So, the next time you turn on a light switch or use wireless devices, remember the young boy from Croatia who dreamed of electrifying the world. Nikola Tesla's electrifying journey reminds us that our dreams have the power to shape the future and bring positive change to the world.

Chapter 10

Niels and the Dancing Electrons

In the cozy city of Copenhagen, Denmark, lived a young boy named Niels. A twinkle of intrigue illuminated his gaze and his mind buzzed with questions about the mysteries of the universe. From a tender age, Niels was captivated by the dance of electrons and the secrets they held within atoms.

Niels loved to explore nature's wonders, from the blooming flowers in the garden to the shimmering waves along the shore. But it was the world of science that truly captured his heart. He believed that understanding the structure of atoms could unlock the key to unravelling the cosmos.

As Niels grew older, his fascination with atoms only deepened. He pursued his studies with vigour, immersing himself in books and experiments. He was particularly drawn to the perplexing behaviour of electrons, those tiny particles that circled the nucleus of an atom like graceful dancers.

Niels ventured to the University of Copenhagen, where he worked alongside renowned scientists. He was introduced to a puzzle that would change the course of his life— the behaviour of electrons in the atom's outer shells. Scientists at the time struggled to explain how these electrons moved and Niels was determined to solve the puzzle.

With creativity and determination, Niels

developed a new model of the atom. He proposed that electrons occupied specific energy levels or orbits around the nucleus, much like the layers of an onion. This "planetary model" of the atom shed light on the behaviour of electrons and their interactions with light.

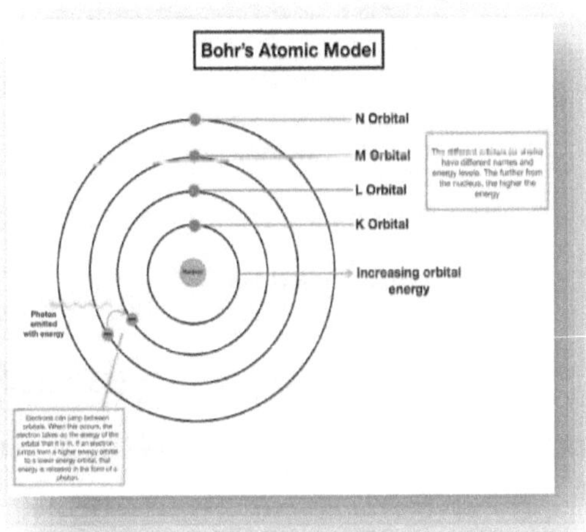

Niels' groundbreaking work earned him the Nobel Prize in Physics, making him one of the most respected scientists of his time. He continued to explore the mysteries of atomic behaviour, developing the "Bohr model" that expanded upon his original ideas.

Niels' model provided the foundation for understanding chemical reactions, the behaviour of elements and even the creation of new materials. His contributions to science were immense and he inspired generations of scientists to study the secrets of the atom.

Niels Bohr's story teaches us that curiosity and a love for exploration can lead to remarkable discoveries. He showed us that even the tiniest particles hold intricate patterns waiting to be uncovered. His legacy as a pioneering physicist continues to inspire young minds to seek answers and push the boundaries of human knowledge.

So, the next time you ponder the mysteries of the universe or explore the world of science, remember the young boy from Denmark who unlocked the dance of electrons and enriched our understanding of the atom. Niels Bohr's journey reminds us that our curiosity has the power to shape the world of science and illuminate the path to discovery.

ABOUT THE AUTHOR

The author of 'Curious Minds' is a Government Servant by profession and a writer by interest. He has publications to his credit in two languages; English and Assamese. He has written many poems and short stories in both the languages. 'My Jungle Book' is his latest publication which is a compilation of fables for kids. He visualizes every child to be exploring and relishing the world of books through which they will in turn navigate through the different aspects of life and universe. In addition to writing, he is also interested in art and photography. His work in art and photography are also published in various exhibitions.